*Dissolution of Matter
Creation of Storm Lightning
and Other Observations*

Eyewitness Report

Dissolution of Matter
Creation of Storm Lightning
and Other Observations

Dennis Bornhöft

© 2012 Dennis Bornhöft
Layout, cover design, production und publication: BoD – Books on Demand
ISBN: 978-3-8448-8965-9

Table of Contents

Preface	7
Introduction	9
The Fogbow with Lens	10
Interpretation of the Fogbow with Lens	15
Cloud with Lens	16
Interpretation of the Cloud with Lens	19
A Particle	20
The Storm Lens – How Storms Arise	21
The Storm Lens in Nature	24
Dissolution of Matter	25
Dissolution of Matter II	30
What Should Be Dissolved?	31
Microwave-Stimulated Image	33
Images that Are Not a Fata Morgana	34
Creation of Rain	36
Method for Creating Rain	36
What I Observed	38
Method for Creating Rain II	39
Miscellaneous	40

A Bright Light	40
A Bright Light II	41
An Orb in the Sky	41
A Small Light	42
A Small Light II	43
Influence on Medical Science?	44
Lines of Force	45
A Gray Orb	45
A Black Image	46
The Image of a Boy	46
A Black Cross	47
A Laser	47
Foreign Spirit?	48
Dreams	49
Epilogue	51

Preface

With this book I would like to draw your attention to observations that I made mainly in the years 2004 and 2005. New physical knowledge can be derived from these perceptions, for example, new knowledge on gravity. However, I can only share what I have seen and experienced as an eye witness. I cannot offer explanations or a theoretical framework for my observations.

I completed secondary school and have been trained as a road engineer and office clerk. Thus, I lack the comprehensive physical education that would allow me to correctly classify my observations. Even so, I have studied the same physical training material used by physics students. During my studies, I was still unable to find evidence of the existence of such events in the physical and technical literature, but I continued my search. You, dear reader, can support my research through your donation. Since my research is more of a layman's collection, your donation is not tax deductible. If you would like to support my research, I would ask you to deposit your donation to the following account with keyword "Donation":

Account holder: Dennis Bornhöft
Bank: DAB Bank
BIN: 701 204 00
Account number: 0041639006
IBAN: DE82701204000041639006
BIC (SWIFT-Code): DABBDEMMXXX

A liability for experiments listed in this book, or a guarantee is not accepted.

I hope you enjoy this book. I also hope that someday you will have the opportunity to make the same observations in nature as I did so that this knowledge is not limited to mere book learning.

Bad Segeberg, June 2009
Dennis Bornhöft

Introduction

This book consists of an eye witness report. It should be read from beginning to end, in other words, from the preface to the epilogue. This is the only way to ensure that it is not misunderstood.

This book centers around the *lens that can dissolve matter.*
Events such as the creation of rain would be impossible to recreate without this lens. It is the power of the lens that produces a variety of wonders.

The reports recorded herein are all true; however, this book only contains a fragment of my observations and does not claim to provide all-encompassing knowledge. It is the first book to describe concrete experiences with elementary particles. Yet, these experiences represent a beginning and not the end of either our era or the elementary particles.

The Fogbow with Lens

In the fall of 2004 I looked out my window and discovered a massive fogbow. The fogbow was approximately three kilometers away and around 1,200 meters high in the sky.

Such an event may not be unusual, but it was just the first irregularity in a series of inconsistencies. The fogbow remained in the same location for almost a week and even withstood a storm. In this case, the storm was the key event. I will simply create a diary in which I describe my memories from the first to the last day.

Day 1

I discover a fogbow approximately three kilometers away from my apartment. It can be seen to the southeast toward the Reinfelder forest. It was around 1,200 meters high.

Days 2 to 4

The fogbow is still there. I take a closer look at it through binoculars. I discover a lens in the upper arc of the fogbow. I understand a lens to be an arc in the air, which could be compared to the lens in a pair of glasses. There are three tubes around the lens, which have the same translucency as the lens. One tube points upward. The other two go to the right and to the left at nine o'clock, twelve o'clock and three o'clock (see image 1). The lens is surrounded by a sort type pf particle, of which there are different particles. Some of the particles are oblong, others are circular. The particles form circles or better said, they seem to move in an intelligent manner.

The movement between the fogbow, lens and tubes actually appears to be very intelligent.

There are no tubes on the bottom side of the lens. Instead, there is a triangle of bright light with the peak at the lens and the base on the ground.

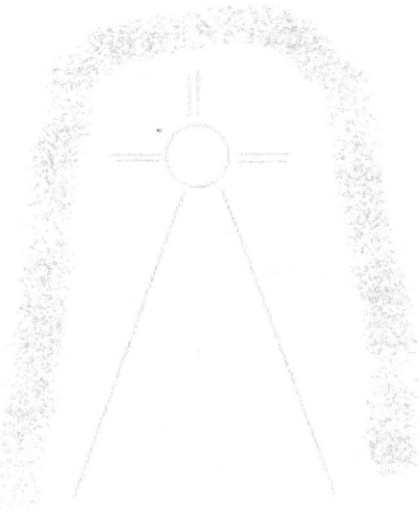

Image 1

The gray band represents the fogbow. The lens can be seen in the upper section with the three tubes and the triangle is shown below the lens. This was the normal state of the fogbow.

Day 5

In the morning I am awaken by a storm. I put on my glasses, look out the window and track the storm. It is not raining. It is light enough to identify the fogbow. The upper half is surrounded by storm clouds; yet, the center of the fogbow is not covered. The lens can be seen. Particles are collecting in the tubes. The particles light up the tubes like lamps and give off a yellow light. The lens trembles. In the sky you can see the formation of a black ray, which is not formed by the above-mentioned particles, but by particles that seem to have come from nowhere. The black ray hits the lens. I can see the beginning of the ray from the corner of my window. It is approximately five kilometers long.

A bolt of lightning discharges that runs within the black ray and only spills out of the ray at the very base. The lightning follows the length of the black ray and hits the lens. **At this point the lens trembles and a type of** *interference arc* **is created.** The lens turns black. It is now much larger than the translucent lens (see image 2). The tubes are no longer visible. As soon as the lens turns pitch black, lightning hits it. I begin counting 21, 22, 23, etc. The lightning remains on the lens for almost seven seconds and then it breaks to the right. The thunder is no more than a crackle.

The black ray is now void of lightning and it shrinks together like a rubber band. It almost seems to me as if black soot were falling out of the black ball in the sky. The lens is no longer black, but is bigger and a greenish color. An orange band can be seen in the middle of the disc (see image 2). The orange band is horizontal and covers a third of the lens. The lens then fades and I decide not to follow the process any further, since my observations have disturbed me too greatly.

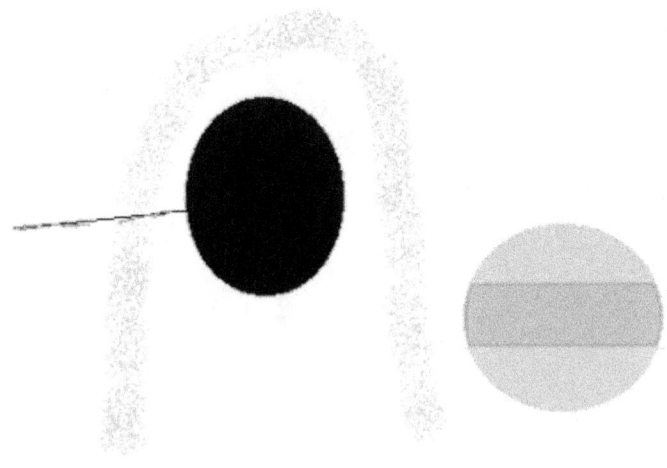

Image 2

This is what the fogbow looked like with the black lens. The important aspect was the transition of the lens and what exactly formed the lens, from a physical point of view.

Day 6

The fogbow is still there, but weaker. It appears as it did on day 2. I seem to have been the only once to have observed the event, since it was not mentioned in the local newspaper.

Day 7

The fogbow can barely be seen.

Day 8

The fogbow has disappeared.

Interpretation of the Fogbow with Lens

I had never seen a fogbow before in my life. However, if you look it up in the dictionary, the term "fogbow" refers to an event seen in optics. The lens, on the other hand, is not a known event nor is this type of particle mentioned in related literature. It is not known that black particles are formed nor that lightning follows the length of the particles. Even the black lens is not known. It could be compared to an interference arc, but cannot be explained by it. The fact is that I observed this with my own eyes.

I cannot explain what I saw. In contrast to other people, I was in the fortunate position to have a front row seat at my window. At least that explains why I was the only one to see something when others did not. To this day, I am still uncertain as to what exactly I saw; however, I can only hope that this book will find a reader who can provide me with more information. However, I have noticed that not very many people take a look at nature, even though there is so much to be discovered. The fogbow was impossible not to see. It would have had to have been directly above me in order for me to miss it.

By the way, I have tried to find a public for this event. I called the ministry of science in Schleswig-Holstein from a phone booth using a calling card, but they told me they were not interested. At that point I did not know what particles were, otherwise, I would have placed more emphasis on the character of the particles. Even so, I knew at that point that it was an event that should not be underestimated. I kept the details to myself as most likely no one would have been interested.

Cloud with Lens

I saw this event in the fall of 2004 and decided to take a closer look at physics.

I went for a walk on a sunny afternoon. I had nothing in particular to do and was thinking to myself, today I am going to see something again. I didn't want to stop thinking along those lines until I had found something. I saw a cloud on the horizon which did not seem to move like the other clouds. This reddish, sunset-like cloud hung in the sky. To be more precise, it stood in front of the neighboring town of Bad Segeberg. I had never seen anything like it.

As I came closer to the cloud, I observed that it really was not moving. It was about two hours before sunset, thus, I had two hours, maybe less, to see everything there was to see.

I went down a street that seemed to lead to the cloud and had a bit of luck. Unfortunately, I did not have my binoculars, but I could see everything very clearly. The street turned to the side and with each step I had a better view of the cloud though the hedge bank. Finally, I came to the end of the bend. From there, I had a perfect view of the cloud. It was located over a field at the edge of the town.

I saw clouds in the shape of a cross that had no depth and were completely flat like a rainbow. One cloud pointed perpendicularly upward, another perpendicularly downward and two others pointed horizontally to the left and to the right. There was a hole in the middle of the cross and there was a lens in the hole with a red border. The inside of the lens looked different than the surrounding air. The lens did not have any tubes. Instead, it looked as if multiple lenses (two or three) were layered on top of each other (see image 3). Particles were coming out of the left cloud. The particles in this cloud were oblong. Around the lens, the particles formed circles. The cloud rumbled each time the particles transformed into circles and went around the lens. This rumbling was a mixture of sound and a slight trembling of the cloud. Through their transformation from lines to circles, the particles seemed to me like living creatures. Moreover, they went between the inner border of the smaller lens to the outer border of the larger lens.

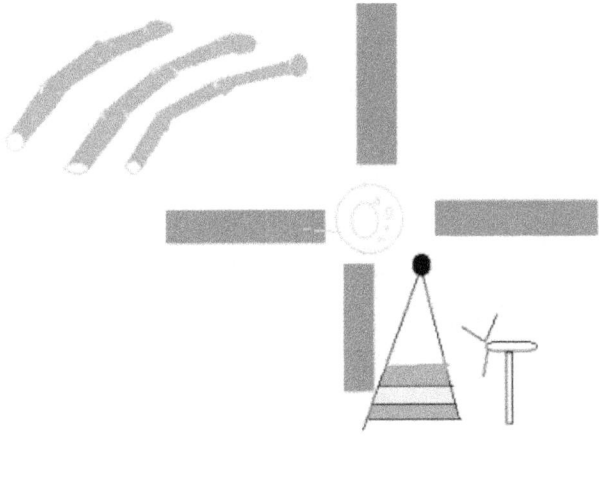

Image 3

The lens was colorless. The particles were white, similar to the fogbow, and the clouds were red.

I also identified around five clouds that also belonged to it. They appeared to be light-blue tubes, parallel to each other to the left of the cross cloud.

What was quite interesting was that the whole thing was taking place in front of a wind turbine. A black dot could be seen in front of the wind turbine from time to time as well as a triangle beneath the dot. The triangle starting at the dot and the base was on the ground. Horizontal beams could be seen in the triangle changing in color from bright yellows to greens. The point and the color beams were as thin as a rainbow.

Interpretation of the Cloud with Lens

What I observed here I had also seen with the fogbow, except that now the colors were similar to those found in a rainbow. I came to the decision that I was dealing with a split rainbow. In other words, all these aspects had to be part of a rainbow as well, but the rainbow had split. In doing so, it revealed that it was much more than the colors of red, green, blue and yellow, there was also a translucency and there was a black dot. In addition, there were white particles, moving in an intelligent manner. If you add the event of the fogbow, rainbows also contain the color gray.

The theory is a little risky, since we are dealing with a rainbow, but if you look at the rainbow with its adjacent rainbow, there is a translucency between the two. If you look at the matte adjacent rainbow, you will also discover the colors of gray, black and white. How a rainbow ends up splitting and assuming forms is a mystery to me. Still, the rainbow is the most visible of my discoveries.

A Particle

This particle flew through my apartment from one window to another. I happened to be sitting at my desk when I heard a rustling behind my back. When I turned around, I saw the particle flying through my apartment. It had two arms and a cavity in which a gray substance stood in the form of a triangle (see image 4). I cannot really explain this particle, **but it was definitely a particle like the ones I observed in** *Fogbow with Lens* **and** *Cloud with Lens*.

Image 4

The particle then disappeared through the other window. I did not touch it although I certainly had the possibility of doing so. When I looked out the window there only remained a black speckle in front of my window into which it must have disappeared.

This particle was seen in *Fogbow with Lens* **and** *Cloud with Lens*. Noteworthy in this case were the two arms and the inside of the circle with the triangle. I do not know what function this particle has.

The Storm Lens – How Storms Arise

I used to think that storms were created by the friction between layers of air. I have now revised my opinion on the creation of storms.

As I mentioned, we are dealing with a storm – in other words, electricity, weather and justifiable human fears. There is a small experiment using a comb and paper shavings. You rub a cotton towel again the comb, the comb takes on an electric charge and attracts the paper shavings. This is why we can assume that electricity is created by friction. I have also studied the physical side of electricity and have to admit that I still don't quite understand it, but the manner in which I describe the creation of electricity is still unknown today.

It was the summer of 2005. The fogbow was part of the past and I had not seen the "cross" cloud again. These were unique events, even though there were still signs of these incidents. Thus, I cannot say what kind of influence they had on the experiments that I performed during that period. I also do not wish to describe these experiments; however, the components of the trials were most likely the actual cause for the creation of the storm lens. I also do not want to know if I could repeat such an experiment. I had set up an experiment and wanted to see what would happen, if anything at all. The fogbow had convinced me that things are not as stable as they seem.

As I said, I had set up an experiment in my apartment and hoped to see something. In reality, nothing happened. I left the set-up as it was and turned to other things. In theory, it can't hurt to leave things as they are, but there was a storm during the night. At first I didn't realize it, but in the end it was right over me. To be on the safe side,

I unplugged everything from the electric sockets in my little loft apartment.

Although I remembered the warning from my childhood to never look at a storm, I watched it anyway. I counted frequent lightning bolts, more than during a usual storm. It struck almost every five seconds, sometimes even three strikes of lightning in five seconds. Thunder followed within one second.

Out of pure boredom, I glanced over at the experiment site. It was only one of many, but suddenly, I discovered a lens in the dark. It had a diameter of around ten centimeters and was slightly aglow. Strips of color went through it.

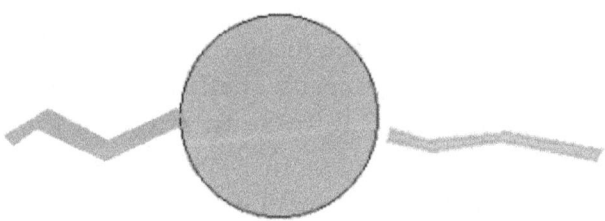

Image 5

This is the "storm lens". Strips of color penetrated it before it uncharged in the form of lightning.

The color strips were also slightly aglow. They looked like little strips from a rainbow. The lens was – just like the color strips – as thin as the rainbow (see image 5).

I watched this phenomenon for a few minutes. The color strips went through the lens one after the other and came from left, right, from above and from below. I discovered shortly thereafter, perhaps one second later, that when a color strip went through the lens lightning, it simultaneously struck in the sky in the same color. If the color strip hit the lens from the left, lightning struck from north to south. If the color strip hit the lens from below, the lightning struck from east to west. Thus, the bolts of lightning were always at a 90° angle to the lens.

The color strips went through the lens and turned 90° until it lit up in the sky.

What surprised me most about the storm was the large number and the wide variety of lightning bolts that it produced. There were bolts of lightning that lit up the sky for 10 seconds as if they couldn't escape. This lightning also rumbled in the sky for ten seconds. Then there was lightning that was horizontal. For instance, there was a bolt of lightning that appeared horizontally two to five times throughout its course. There were also bolts of lightning in all colors - green, red, orange-red, etc. The lightning was 50 meters to a kilometer long.

A few days later no signs of the storm could be seen, but it was still humid. I thought I saw a ball of lightning hanging dangerously over a settlement, but it was actually a round plane of electricity. This lightning "ball" stayed there for almost a half an hour and made no noise and a triangle could be seen below it containing horizontal red and white beams. The ball was too far away to determine its exact size, but it may have had a diameter of possibly 70 centimeters. I am pretty sure that this was electricity.

The Storm Lens in Nature

I actually still (in the year 2009) see a lens, such as a car on a rarely driven street, streak through my apartment from time to time. This is very rare, but I occasionally discover a few through pure chance. Even if this lens has no property in my eyes, I am convinced that there are many similar lenses on the earth. These lenses probably not only exist on the earth and in the skies, but probably at the earth's core as well.

The ones at the earth's core are probably lenses that could create heat. Like the storm lens that attracts strips of color without knowing where they come from, there must be a similar behavior with lenses in the earth's core and in the northern lights that form color strips, thus creating a form of sustenance for the lenses.

This assumption would create a new theory about the high temperatures at the earth's core.

Dissolution of Matter

The discovery of the lens came quite late. At first, I did not know what the term "lens" meant. To help clarify, I should refer to an event that suggests the beginning of the lens. I am not sure how long it had been hanging on the wall. To be more precise, it was at an angle on the inside of the ceiling.

A few months before my discovery, I had a bucket full of stones in a corner of my room. In addition to fist-sized stones, the bucket also contained water and a few new matches. In any case, one evening I discovered a black spot on the wall where the bucket was. This spot probably consisted of the same substance as the black ray in which the lightning ran (see Fogbow with Lens). The spot had a diameter of 20 centimeters, but it wasn't really a spot, more of a kind of muon spot. I say "muon spot" since no other particle can come into question. You could have reached through the spot and not have touched anything real. I also had a television on the cabinet, as well as some keys with sand and matches. In any case, I was able to dissolve the spot by fooling around with two magnets. Suddenly, a substance came out of another lens directly above the bucket containing the stones. It was similar to the some of the sparkling minerals on the stones.

The lens dissolved, or better said the blackness dissolved. I was able to catch the substance that came out of the blackness in my hand. As I said, this is the same activity I observed with the fogbow only there was one difference: Small lightning flashed following the dispersal of the substance. This lightning was maybe ten to 20 centimeters long and was actually very cute. It made a noise like lightning. It was a humid evening and I thought to myself, "Why not? Why aren't there such small storms more often?" Why there are only big storms totally escapes me.

I recognized at this point, that we have not gotten very far with electricity, at least in regard to our knowledge about electricity, but it seems to be that smaller lightning cannot strike because of the black substance. By the way, there was a debate between George Bush and another presidential candidate on the television at that time. I thought, "These are both very important events" and jumped for joy.

This is the history of the lens. It was steadily located on my wall for over half a year. Later, as I was setting up my experiment site, I realized that an experiment site must have been responsible for the lens. It was a glass full of magnetic bars from a magnet game. These bars were made of plastic and had a small magnet on the end (see image 6). Unfortunately, I had to throw the experiment away, since I did not have the same success with the new materials. As I said, the experiment must be taken in context with the history.

Image 6

Magnetic bars. The lens disappeared when I removed this glass.

Now, how this is related to the lens. The lens had a diameter of approximately ten centimeters. It was translucent and a type of translucent tube was attached to it that extended beyond the roof. One day I saw a spider crawling toward the lens. I thought this would certainly be interesting. I myself had never touched the lens. The spider was not very big. It moved toward the lens and stopped at the edge, to the right of the lens. Suddenly, the spider was gone and only to reappear a few seconds later on the left side of the lens. Then something happened that I had not expected. Blue-white beams of color appeared and took the spider with them - probably through the tube (see image 7). I cannot really say whether or not the spider was actually dissolved by the tube. In any case, the spider had disappeared. It had dissolved. It was gone forever. If asked about the whereabouts of the spider I can only say that I later saw these black beams, through which the lightning usually went, more frequently than before. The spider was in these beams. I do not know if it crawled into them. In any event, the spider moved with the beams which disappeared as fast as they had appeared. The spider was now very flat, as flat as the rainbow.

Image 7

I am a smoker and my room is frequently full of cigarette smoke. What I did not really understand was that the cigarette smoke was also dissolved. Sometimes I saw a cloud of smoke in the middle of the room that I could not really explain. These smoke clouds were in the room, but could not be touched or dispersed. I simply determined that the collisions became more and more infrequent with the increase of time after its dissolution.

Later I saw yet another lens on the desk in my room. This also worth mentioning. What came out of the lens was not a stony substance, but a living mosquito that once it had developed from a black spot flew happily away. The lens was mobile. It moved back and forth and at times it was right in front of my face. It was here that I also observed the development of the mosquito out of the black spot in the middle of the lens. This all happened within a few seconds. It was basically no more than three seconds.

I could safely say that the lens that dissolved the matter was in my apartment for over half a year. During this time it stayed in the same location and never changed its position.

Dissolution of Matter II

I often went for walks, sometimes also at night on the Kalkberg, a cliff in Bad Segeberg. There is a lookout platform at the top of the cliff from which you can see parts of the city. This is especially interesting as the ascent is very difficult.

During one of my night hikes on the mountain, I turned around on the platform and saw white-blue beams of color to the East. The peak then shuddered. I personally even believe that the whole mountain shuddered.

I see this as proof that we are not that far from *graviton*. Graviton is the elementary particle of gravity. The white-blue strips assumed the contours of my body; however, they were much larger. The small of my back would have had to be three meters wide. The white-blue beams of color were maybe five meters away from me and hung right above the center of the stadium in Bad Segeberg.

Unfortunately, I only caught a brief glimpse of this image. What I also noticed was the fact that this image seemed to disappear when I turned around and looked at it directly – as if these phenomena were not immune to being looked at with the eyes.

Apparently, I had carried parts of this technology directly on my body. This was another reason for me to set up my experiment site and not to confront the public it. After this event, I once stood in front of the mirror and saw white-blue beams in my eyes. I could still see very well and was not limited by them at all. What remains are the tremors.

Even if this only happened a single time, I remember it very well.

What Should Be Dissolved?

If we someday manage to create such lenses, may even in mass, then the question arises as to what should be dissolved. In reality, anything could be dissolved – metals as well as semi-conductors or other chemical elements. Even energies could probably be dissolved.

Yet the question remains as to what should be dissolved – for example, atomic waste among other things. I would like to start a discussion about this. I would personally prefer if atomic waste were not dissolved, but instead dying people.

I would like to briefly explain why I think that people could continue to live.

I lived in my apartment with the lens and gained a few experiences with it and its energies. Thus, I would have to attribute it to the lens that one day my feces was white. As far as I know, the dark color of feces is caused when red blood platelets die, thereby giving the feces a brownish color. Thus, I am of the opinion that my feces had no color because my blood platelets did not die that day. That is why I am also of the opinion that the human could continue to live, even if in a world that would hardly have anything, or much less, to do with our world.

I even think I know what it would feel like to dissolve. In the end that was also the decisive reason I disassembled my experiment site. I walked through the pedestrian zone of Bad Segeberg. I suddenly felt as if I were a giant crystal that was being beaten with a hammer. I wasn't shaking, but I shook like a crystal that was in danger of being shattered into tiny pieces.

Another time I was in a neighboring supermarket. Repairs were being made to the buried cable of the shopping cart shed. I stood at the checkout and was about to pay, not even five meters away from the construction. Suddenly, I started to pulse, like the vibration setting on a cell phone, just much faster. My body functions, however, did not pulse. It is particles that create this kind of effect.

Microwave-Stimulated Image

I have made a number of experiments, including the *Microwave-Stimulated Image*. I set up an experiment site in my microwave and turned the appliance on. Nothing really happened beyond a few sparks in the microwave; however, when I looked out my window I saw a gray circle a few hundred meters away with a diameter of maybe 20 meters.

I am mentioning this experiment since there can be a connection between an experiment and the distance. First, however, I would like to describe the experiment. Perhaps the result can be reproduced after all.

I used a glass microwave bowl and covered the bottom with sand. Then, I folded a square of aluminum foil into a triangle by turning in the corners. I placed the aluminum triangle in the round microwave bowl. The corners touched the edge of the bowl perfectly. I removed the turn table from the microwave and I put the bowl in the microwave. Then I turned on the microwave. A spark came out of one of the corners of the aluminum foil, which cracked the glass microwave bowl. To be more precise, the spark made a round hole in the bowl with a diameter of two centimeters. The hole in the bowl was at a 90° angle to the gray spot that formed in the sky.

From this experiment, conclude that the image in the sky was caused by the microwave experiment: *A microwave-stimulated image.*

Images that Are Not a Fata Morgana

I would not have mentioned the microwave-stimulated image had it not been for further events that evoked images. There is not a lot to tell in this case either. In the beginning I recognized red spots in the evening sky. I wondered what the spots were, but I could not see the other side of the spots. Then the following occurred. I was sitting at my kitchen table and had the television on top of the cupboard about three meters away. Suddenly, a second image appeared in front of the screen. This image was as thin as the rainbow. I could see how it dissolved, not like a television image, but like a portrait where the colors of the image had been torn apart. The size of the image was about 70 centimeters diagonally. It was black with a thick green frame and a thin red frame on the very outside. I actually did not pay any special attention to the image. Much more impressive was the fact that I had no idea as to how this image had formed without a larger experimental site.

A few days later I had a gold-silver emergency blanket lying on my desk in its original packaging. Next to it was a microwave bowl made of plastic filled with gold-colored minerals. I had brushed the minerals off stones using a steel brush. As it became dark toward evening, I looked out my window and saw a screen that was 40 meters long and 20 meters high. The image that I saw was an image that I had just been imagining in my mind, but what was not in my mind was the golden base color that the image exhibited. The image – also as thin as the rainbow – lingered approximately five seconds. None of my neighbors seemed to have seen it.

I saw such images a few times after that. For example, I once observed how a colored laser formed between two buildings in the city center of

Bad Segeberg. A screen that was 15 meters wide and four meters high formed above the color beam for five seconds. However, if you looked at the screens directly, they disappeared.

Creation of Rain

I have designed a few techniques that I truly believe have the potential to create a rain shower. However, this method has never been tested in the desert and from the looks of things, I doubt that I will ever be able to visit the desert. I used to think that clouds were formed when water was heated, and who knows, maybe it is true that clouds come from heated water and then the rain falls from them. In any event, my technique is not a method in which you see a single cloud that you could get to drop rain. Rather, my method aims at creating a weather condition. The weather will come to the location where the experimental site has been installed. I will describe how my experiment progressed and what I observed.

Method for Creating Rain

Take a plastic Soda-Club fruit syrup bottle. Empty and clean the bottle, removing the foil. Next, fill it with gold-colored minerals that are the size of grains of sand. Add enough water to the bottle so that a small puddle forms on the sand. The bottle is then closed, shaken well and wrapped in aluminum foil (see image 8). The aluminum foil is folded in at the ends of the bottle. Place the bottle on a wooden floor in the attic. Wrap the cord from an electrical appliance, perhaps from a lamp, around the bottle three times (220 volt).

You will then need a dome made of aluminum foil. I then wrapped aluminum foil around a set of books, which formed a type of pipe. The pipe should be smaller at the top than the side that is on the floor. The dome is then placed over the bottle with the cord, leaving sufficient space on all sides of the bottle. The experiment is now done and should be left for as long as the cloud formation needs to get from the ocean to your area – generally seven to ten days.

Image 8

Aluminum foil is wrapped around a bottle. A pipe is created using aluminum foil.

What I Observed

I could see directly out of the attic window when I was setting up the bottle. During the experiment I also observed that a hole was formed in the clouds above my window, right above the experiment site, through which blue sky could be seen.

Moreover, it was possible to use the window as a mirror in the evening and small glowing points could be seen on the bottle, which I call *virtual photons*. A light should definitely be on in the room. These glowing points can only be seen in the reflection in the window.

When the weather with clouds arrives in the town, a lens is formed in the sky with a diameter of multiple kilometers around which particles whizzed.

If you set up this experiment and the next day is foggy, you can count on rain arriving seven days later and lasting a few days.

Method for Creating Rain II

You should only use this method if you could not create rain using the first. It is very simple.

Take metal candy tins (see image 9). Empty and clean the tins. Then place four tins on top of each other and place them in a cupboard. Stack three more tins and place them next to the four tins. It should rain before you disassemble the experiment.

Image 9

Miscellaneous

In this chapter I would like to share the observations I made, but I cannot classify. In most cases I simply cannot find further insight. By no means does this mean that I should keep these perceptions to myself. They should definitely be mentioned so that you know: This happened on our earth.

A Bright Light

I had made myself a pasta salad. The bowl was on the stovetop, which was turned off. There was a ladle in the bowl and the whole thing was covered with aluminum foil. Then something happened that I had not counted on. It was dark and I was lying in my bed. Suddenly, a circle appeared at my feet that shone brighter than the sun. It had a diameter of around 1.2 meters. Tubes were coming out of the circle that seemed to touch my feet. At least that is what it looked like. The circle shone for maybe ten seconds, then disappeared again. First I thought that it might have something to do with my pasta salad; however, I later remembered that I had already seen this light before directly in front of my small window. At that time, the light lit up the entire backyard. This one was just as bright. It did not dissolve like normal light, but the way that paint breaks off in little pieces. Moreover, it seemed as if the light was retracted by the source of the light.

I frequently saw smaller versions of this light until sometime in 2007. It had a diameter of possibly 30 centimeters. This light also had tubes and touched my body without my being able to feel it. It was more that I saw this small light and watched the event take place.

A Bright Light II

At the beginning of 2009 there was sheet lightning in Scandinavia and Northern Germany, which was also recorded by the press. A meteorite crashed and caused the sky to glow blue for a few seconds. Unfortunately, I was in bed and could only see the glow out of the corner of my eye. Since the press reported this event it is not actually a topic for this book, but I had also seen sheet lightning in 2004 that was not that different from this one.

At that time I had set up an experiment that I do not want to talk about. I used my voice to call into the experiment and a red glow appeared in the sky south of Bad Segeberg at an estimated distance of 20 km. For a few seconds the earth looked as if it were covered in red light, like a submarine. The glow was not the only thing I noticed. I could also see black images that seemed to be made of soot. They were at house level and had a diameter of one meter similar to oversized cornstalks. At first I thought they must look the same all over the earth and that these oversized cornstalks probably played a not insignificant role, such as gravity (i.e. the earth's gravitational force). This, of course, was only an assumption.

An Orb in the Sky

This is probably the worst event I can remember. It happened in the evening. I actually watch the stars quite often. Even though I do not have a problem with my eyes, I frequently saw the stars not as round forms, but directly, like color lasers. The stars were not round. Instead, they looked like streaks. Of course, stars are round, but it seemed that each star in this galaxy was specifically reconstructed in the numerous layers of the sky that Earth has. The structure of the image of a star was none other than a laser beam. Sometimes I even had the impression that a laser was again issued from this image that united with other lasers.

But now to the actual event. I saw an orb in the sky that curved the sky to such an extent that the stars seemed to move. Looking at the orb through binoculars, a laser could be seen coming from this orb that sought out a building, which then shook. I still don't know today where this orb came from or what it meant, but the sky must contain a few things that we still do not understand.

A Small Light

Once, as I was riding my bike through the area one evening, I passed by Bad Segeberg's water treatment plant. There I saw a light with a diameter of perhaps three meters right on top of a rod to the left of the treatment plant. This image was as thin as the rainbow, yet it shone so bright, I immediately identified it as an event. I watched this light for quite a while and then wanted to find out if I could see it from my apartment. So I rode my bike home. I got to thinking again, this time, about the moon. It was at least a hundred times larger than normal so that I was even worried it might collide with the earth. As soon as I got home I looked for the light. I could not see it from my window. Then I saw the moon at its correct size, which comforted me greatly. I did see a light, but I do not know how the light was created. Then I saw the moon, as if through a lens, bigger than it is and much earlier than usual, namely before the actual moonrise. To this day, I can only explain these occurrences with the idea that the sky must have curved to such an extent that the moon seemed a hundred times larger. What causes this kind of curving, I do not know, however, the light was not a star that had been magnified a hundred times. I stand by my assumption that it was caused by something else.

A Small Light II

I often look out my window at night. When I do, I often notice lights appearing that I can only explain by saying they are star constellations. To me, they appeared to be constellations that had been projected onto the earth. These constellations had diameter of a few meters. I am not familiar with constellations, but these were constellations. They were not spatial, but rather flat discs composed of glowing particles. The unique aspect, however, was that they could only be seen from my apartment.

When I rode my bike toward the images they were no longer to be seen, but, back in my apartment, they were still discernible. I only observed these appearances on a few days. Still, I find it significant that the sky is capable of creating such refractions, thereby creating this kind of magnifications.

The most significant discovery though was a pyramid of glowing, white particles. Do such constellations actually exist in the universe and no one has discovered them yet, since we can only see the base of the pyramid from Earth, and is it possible that these constellations are only made up of particles?

Influence on Medical Science?

I am cleaning my toilet with a steel brush. It was stubborn grime on which I had already used a variety of cleaning agents. I was bent over the toilet bowl. I had a steel brush in the left hand and my right hand was resting on the toilet bowl. Suddenly, I saw a colored spot surrounding my body. It was a red and green spot that was as thin as the rainbow. They were diamond shaped and formed a ring around my stomach. The spots were at a distance of around 20 centimeters from my body. All in all there were only four spots.

When I turned around and looked out the other attic window, I also saw a roundish white spot. This spot was maybe five meters away from my body and was different than the colored diamonds.

I asked myself whether the colored spots did not signify organs, which performed part of their task outside the body. Their discovery was reserved for me. In the experiences I have had, everything leads back to a rainbow. If you take a closer look at a human's internal organs, you can actually find all colors of the rainbow within the organs. Thereby, the colored diamonds were colors that could only come from organs. I am convinced that this is proof enough that the rainbow is part of human health. When you use the human body as an example, the clearness of the translucency plays an important role. In the same way, some of the human body is transparent as well. The question that arises is what purpose do the translucency and clearness have, and not only in the human body, but also in a rainbow.

Lines of Force

In my observations I can only ever refer to that fact that I saw them with my own eyes. I was fully aware that risks were involved, even though I did not know what those risks were. It is the same with the lines of force that I saw before my eyes time and again. They consisted of small gold and black diamonds that were lined up one after the other, thereby forming strands multiple meters long. The strands were not even a half centimeter thick yet I was still able to identify the structure. Once I saw a strand weave right through my apartment ending at the towel in the bathroom. When I then looked out the window, I saw a white bolt of lightning of monstrous breadth that struck near the neighbor's house. I am sure that this lightning was related to the strand.

Another time I identified a strand of fog multiple meters wide right in front of my window. When I took a closer look at this short event, I realized that the strand of fog was dissolving and ended in a line for a brief moment. It then returned to a strand of fog. Finally, it disappeared altogether.

A Gray Orb

In *Cloud with Lens* **I mentioned a discovery that I couldn't really classify.** I saw a small gray orb with a diameter of perhaps ten centimeters hopping across a field. This orb was a foggy gray and had a kind of shape, even though it actually consisted of steam and did not really dissolve. On the other hand, maybe it was only an unknown part of a rainbow.

I only saw this orb twice. The second time I saw it in the fall of 2005. I saw a couple on the street, made up of two people. I saw a gray orb between and above their heads. Black beams came out of the heads that hit the gray orb. I am still not sure what the orb meant. I only know that sparks ran along the black beams. The orb was only visible for a few seconds.

A Black Image

One day as I was leaving my apartment and was standing at the exit I saw a black image form on my right arm between the lower and upper arm. It assumed the shape of a triangle. The image was black as night and fully impenetrable. It seemed to be a type of sail, since it puffed itself up. I am not sure where this black image came from, but it unsettled me somewhat. After some back and forth, it finally disappeared when I wasn't looking. **I am not sure what caused the image, but I am sure that it was during the period when the** *dissolution-of-matter lens* **was active.** It must have been an appendix that was not as easy to get rid of.

The Image of a Boy

I woke up one morning and already knew something wasn't right. **The** *dissolution-of-matter lens* **was still active.** What I saw when I opened my eyes was the image of a boy in the middle of a milky bubble. The image was definitely three-dimensional. The bubble went through my arm so that I was hugging the boy with one arm. What was interesting about the image was that I am not sure if it was merely an image or possibly the creation of matter. I have already mentioned that mosquitoes appeared in front of my eyes, seemingly out of nowhere, out of almost nothing, but actually from a lens. Maybe this bubble was a form of lens, a type of spirit that can create matter. I am not quite sure where the spirit came from. **Maybe it was a combination of all the buckets I had in my room, and the** *dissolution-of-matter lens.* This bubble could not have come from nowhere.

A Black Cross

In the middle of a normal day, I saw a black cross that hopped over my power cable like a scissors. This was the only reason that I examined the cross more closely. Since it hopped like a scissors, it made a noise, which could be compared with the sound electricity makes when its charge waxes and wanes. As I said, this black cross hopped over a power cable. The only other thing I can say is that this cross was made of photons or whatever. The fact that photons can may that kind of noise makes me doubt the soundlessness of photons.

A Laser

I described the *particle* **as I experienced and saw it.** This particle was a gray triangle. I have absolutely no idea what the purpose of the triangle was, but what I saw should not go unmentioned. I was having my lunch in the kitchen and was looking straight ahead when I discovered a gray substance on my broken CD player on the other side of my room. This gray substance was diamond-shaped and rested on the CD player. I looked at the gray substance and was not sure if it might not be dangerous. I had never seen this substance by itself.

Suddenly, a laser came out of the substance that hit me on the forehead, right between the eyes. Moreover, the laser hit a zit on my forehead. It was orange colored and maybe a quarter of a centimeter thick. Still, I felt the impact on my forehead, it tickled a little. The laser itself was not a simple laser beam. It was more of an orange band. The fact that I could feel it is reason enough for me to say that it was not destructive, but still noticeable.

Foreign Spirit?

In the observations I have made the question keeps arising as to a foreign intelligence. A spirit would be required to create a mosquito using a lens, as I described it. I would like to share a few observations that I have made to make it clear that I assume there are foreign spirits. For example, I have heard voices twice. When I looked for the source of the voice I saw two lines composed of black particles in the air from which the sound came. I can only remember one word: "Sometimes I could" said the voice. The voice was feminine and very disciplined as if the female to whom the voice belonged sat working at delicate needlework.

The other observation I made was a screen, as I have described a few times. Latin letters could be seen on the screen that was composed of particles and colors.

I saw this at least twice. As hard as I tried I was unable to read the writing. The writing was written in letters that were five centimeters large. Despite this fact I was unable to read them and their meaning remains hidden.

I have not written down all of the observations I have made.

Dreams

I do not know how it is for you when you lay down to sleep and close your eyes. In my case at least I saw images after this event.

Unfortunately, my imagination does not allow me to create my own images. I let everything come to me. In my case, I see images as if looking through eyes. I do not know whose eyes they are. I only know that it is actually being seen, as if by a spirit.

Sometimes I see the earth at least 300 kilometer below me. This is already space.

Other times I see images that I cannot classify at all. Thus, one time I thought I saw an older teddy bear; however, when I was watching television a few days later, I became sure that it was the prison door of Natascha Kampusch. This woman had survived numerous years in a fall-out shelter. I didn't understand this image at all. Not until I watched television did I understand what I was dealing with.

Another time I saw a scene that I didn't understand either. It was kind of a combination of sex and naked people, of generals and soldiers, fencing on a table. The next day I learned that one of my old classmates had died on his farm from an accident on the job. The police, fire department and emergency doctor were on site. I am sure that these are images that have seen the eyes of a spirit. These images have their own character.

None of this has any conclusiveness. However, when I see a spiritual image with such a spiritual character, in which black and white static appears as we see on the television when no program is running (anti-matter), I am sure that this is not just a figment of my imagination.

It actually seems to be that some people only have to close their eyes to make contact with another existence.

I consider it to be good news that not everything we see comes from our brain, but from another spirit. That is also why I wrote this book.

Epilogue

I have now recounted a whole series of examples. Even though I may not be a very good author, I think the attentive reader can imagine what I have seen. This book serves no other purpose. Except for the method for creating rain, which I do not know if it works or not, this book has fulfilled its purpose - namely, to ask a question. The question is whether what I have reported here, as an eye witness, is credible? I am not sure. I can only say that I have seen these things and am glad not to see them any more. I would be pleased if my perceptions and this report were taken seriously and, if perhaps someday, research were done along these lines.

www.ingramcontent.com/pod-product-compliance
Lightning Source LLC
Chambersburg PA
CBHW050245230526

45470CB00005B/2118